Praise for David Evans's

Achieving ZERO

"With clarity and broad understanding, the founder of David Evans and Associates reflects on a half century as a successful engineer-businessman. He recalls both the trials and fun of building the reputable engineering firm which bears his name, and reveals the instincts and insights that have served him well. A fast, satisfying read on the business and human side of consulting engineering. Both students and current practitioners can gain from this enjoyable book."

David A. Raymond
President, American Council
of Engineering Companies

"This is an incredibly helpful book for graduating engineers or young engineers in consulting firms. It's a delightful story of how one young engineer evolved, step by step, until he'd built one of the most prestigious consulting engineering firms in the country. What's most unique, David Evans has the rare gift of making serious commitments to people he meets whom he believes have both talent and passion for their pursuits. And he built a successful firm of several hundred around his belief in people. A great lesson in entrepreneurship and leadership."

Stuart W. Rose, PhD
Professional Development Resources Inc.

"No matter where you are on your vocational or professional path, read and use Dave's insightful book. It is filled with wisdom aplenty for all of us. Whether you are an aspiring engineer, rising employee, potential entrepreneur, or firm executive, you can learn from Dave's pilgrimage: career planning, firm organization and culture, decision-making, leveraging successes, and learning from every resource life offers. The engineering profession has just cause for great pride in its contributions and achievements to society. These benefits come from firms like Dave's that commit to unselfishly serve their communities' needs and embrace that mission with zeal, courage, integrity, and humanity. Best of all, stewards like Dave make sure that commitment stays true, even when they must transition the leadership to a trusted elder. David has captured his wisdom for our benefit; read this book, use it, and let Dave know how your own pilgrimage goes."

Louis L. Marines, Hon. AIA
Founder, Advanced Management Institute
for Architecture and Engineering
CEO Emeritus, American Institute of Architects

"I would like for every engineering student and young engineer, especially those who are thinking about starting a firm of their own, to read David's book and gain some valuable perspective. I firmly believe that he, and engineers as a profession, have completed lasting projects that achieve much more than ZERO. The work environment David put together in the firm—with all the values, collegiality, and pride I saw in the DEA employees—created a culture where people *wanted* to come to work every day rather than *having* to come to work. That kind of workplace and the high values it embodies means much more than ZERO to many, if not all, on the DEA team. In my opinion, David should add a 1 in front of the ZERO, because his firm has become a 10! The experiences shared in *Achieving Zero* should encourage all young engineers to follow their dreams!"

Frieder Seible, Dean
Jacobs School of Engineering
University of California, San Diego

"Achieving Zero is an excellent history for understanding the development and growth of an organization. David Evans gives us the opportunity to learn from his experiences, both positive and negative, and it's fun being along for the ride with him and the firm from the beginning to the present. This is a great book for those who want to look at history before creating their own."

Steven J. Isaacs, PE, Assoc. AIA
Division Manager, FMI Corporation

"Any guy who would have a fleet of classic cars as company vehicles instead of Honda Accords can't be all bad! I have known David Evans since the early 1980s, and he is not a stereotypical engineer with a short-sleeve white dress shirt and pocket full of pens. He is a real leader who has made his mark on the engineering industry. He started his own firm in 1976 that today employs more than 650 people. He has had an amazing career by any standard. In his book, *Achieving Zero*, David Evans tells the tale of his own life as an engineer and how he built his consulting engineering business with humor, grace, and humility. Anyone in the engineering business would enjoy reading *Achieving Zero*, as I know I did."

Mark C. Zweig
Founder and CEO, ZweigWhite

"An honest, open appraisal of the ups and downs and philosophy of Dave Evans in creating a firm and keeping his employees focused on the greater good. The civil engineering profession has changed substantially since David Evans first began his career. This text provides an important historical and cultural context for upcoming civil engineers, who need to understand their cultural heritage. I will certainly share Dave's advice with upcoming Civil and Environmental Engineering students at Portland State University as they become part of the civil engineering family!"

Scott A. Wells, PhD, PE
Chair and Professor, Department of Civil and Environmental Engineering
Portland State University

"*Achieving Zero* tells the valuable story of David Evans's authentic leadership in founding a super firm. This is both a personal and worthwhile story by a legendary professional."

David Aitken, BArch, FRAIC
Aitken Leadership Group

Achieving ZERO

David F. Evans, PE, PLS, FASCE

Copyright © 2013 David F. Evans
All rights reserved.

ISBN: 1-4802-8428-9
ISBN-13: 978-1-4802-8428-9

To all engineers, young and not-so-young, who make decisions every day to ensure we have safe, healthy and enjoyable lives. You make the world go 'round.

Contents

Acknowledgments i

1 Zero: The Top Score! 1

2 Deciding to Become a Consulting
  Engineer 5

3 A Firm Is Born 17

4 Bumps in the Road 27

5 Some Surprises Are Good 31

6 Defining the Firm 37

7 Stepping into the Computer Age 45

8 Clarity of Purpose 53

9 Building a Board of Directors 69

10 An Article Published by
   Civil Engineering News, November 1996 75

11	Understanding Our Plight	89
12	People, People, People	95
13	Enjoyment Supports Creativity	109
14	Transitions	117
15	Last Letter to the Firm	123
16	Advice to Prospective Consulting Engineers	127

Further Reading 133

Song Lyrics and Excerpts 135

About the Author 141

Endnotes 147

Acknowledgments

While this tale is about my exploits as an engineer in the consulting business, I cannot proceed without clearly and extensively thanking my wife, Marcella. I won't put much of her working life at the company in this book, but from the first day, when she came to the new office with me to provide secretarial, receptionist, and moral support, she has provided unfailing, invaluable support, advice, and help. And, of course, she was unpaid!

While Marcella knows of my incredible appreciation for her efforts, no words could be penned without formally acknowledging them. Thanks, Honey.

I also must acknowledge David H. Gould. Dave agreed to leave his secure role at the firm we were working at as young engineers to cast in his lot as my helper, designer, draftsman, and friend. Dave is an incredible design engineer. While not a licensed professional, he designed complex projects and never, ever, had to redo, fix, or change his efforts. His humor also appears in this book.

I do not type. All work here has been penned by my hand. I should have been a doctor with this handwriting of mine. But my executive assistant for more years than either she or I can remember has deciphered my scribbles. Her support and abilities were always appreciated by me, and she deserves much gratitude from me. Thank you, Mary Beth Hernandez.

Making my words and writing come across in the King's English required the knowledge of a professional. My sincere thanks and admiration for her skills go to Karen Newcombe, who edited, formatted, and commented on the text, and gave me valuable advice.

Chapter 1
Zero: The Top Score!

"The game commences, for the
usual fee, plus expenses."
– Dire Straits

 I am an engineer, actually a consulting civil engineer with professional licenses in five states, also an entrepreneur, a professional land surveyor, and the founder—thirty-six years ago—of a major design firm. Today I am chairman emeritus of the firm, and I am still on the board of directors.

 For thirty-five years I wrote an encouraging paragraph or two every month to be sent to all the employees—up to one thousand at the height of the firm's growth. I also penned several magazine articles and was quoted in or added short pieces to multiple stories about the design and construction industry. During some of my many presentations at industry conferences, people came up and urged me to write a book. My wife did too. So this is it. Some facts, some fantasy, some parts penned by others that seem appropriate, with appropriate credits, and, I hope, with significant humor. So begins my story of Achieving Zero.

Consulting engineers and their design firms are selected by their clients on the basis of how well they can convince the clients that they have the

- Best experience,
- Best approach,
- Best skills and talent,
- Best program,
- Best proposal,
- Best relationship with the client, and
- Best staff members for the client's needs.

The consulting engineer virtually promises to make the required miracles happen, to meet and exceed all expectations, and to do so on time and on budget.

Upon doing so, no significant recognition of this success is expected from the client. You have only accomplished what you proposed. You have achieved a Zero. Doing less than you have proposed, well, you go down from there.

In spite of this, engineering is a marvelous profession. President Herbert Hoover, in his "Addresses Upon the American Road," said of engineers,

> *His is a profession where he has the fascination of watching a figment of the imagination emerge through the aid of the sciences to a plan on paper. Then it moves to realization in stone, or metal, or energy. Then it brings jobs to men. Then it adds more and better homes. Thus it spreads progress and opportunity over the land. That is the engineer's high privilege among professions.*

Achieving Zero

While being satisfied with achieving zero may sound discouraging, it is not. Finishing a design, seeing it grow into reality, and then seeing the results being used by society is a very fulfilling reward. Recognition by the client and peers is nice, but not expected. Becoming a consulting engineer requires traveling a specific course, overcoming obstacles, and reaching mileposts. My travels along that path are the essence of this book. My hope is that a young beginning engineer reading this book will find appeal in the joys I found in being a consulting engineer and will make becoming one a very worthy pursuit.

Chapter 2
Deciding to Become a Consulting Engineer

"My, but we learn so slow, and heroes they come and they go."
— The Eagles

While this epistle is mostly about the firms created under my leadership and my tenure as president, CEO, and chairman of the David Evans entities, I should at least begin with establishing the setting: How did I get here?

Upon finishing engineering college and passing the engineer-in-training exam (now the Fundamentals of Engineering exam) as part of that process, I became a beginner engineer. While 1961 was not a banner year to begin an engineering career, I found myself gainfully employed on January 30, 1961.

My first day on the job at Pacific Gas and Electric Company (PG&E) in San Francisco almost became my only day at PG&E. It seems my wife of sixteen months decided January 31 at 3:00 a.m. was the exact right time to bless me with a beautiful, healthy daughter. Getting to work after this early morning event at the hospital required a streetcar commute. While I ended up on a car later than I had planned, the San Francisco Municipal Railway system saved my job by causing delays for all on board.

So, on day two, I arrived about thirty minutes late to the frowns of the four executives in the small office. I mentioned the streetcar problems, and at least one of the men was aware of it already, so I was excused from further discipline, but still, frowns. At noon, I purchased cigars to hand out in the office and announced my new fatherhood of that morning. Even the frowns disappeared.

One of the interesting things about an engineering education (at that time) was that you learned a ton of technical stuff: how to design, calculate, formulate, tabulate, and draw engineering solutions to the problems the profession presents. However, there was no education about what you might or should expect to do in an actual job.

I was filled with visions of designing structures, bridges, roads, and related infrastructure elements, and I was confident in my learning. It would have been very helpful for someone in academia to tell this new graduate, "The industry will give you no responsibility because you have no experience." Learning this firsthand was a big shock to my ego and the high expectations of all this newly minted engineer would do. As it turned out, my place at PG&E was totally unrelated to designing anything. I was assigned to report the status of weather and progress at various project sites to the construction managers. Design was in another section of PG&E that was not hiring.

I was sure the problem was the job and not my lack of experience. I had applied for and had been offered a different job—a junior civil engineer position with the City of San Mateo. My interview with the city engineer impressed me with the wonderful projects going on and coming up in the city—things that needed design. (It also meant more money for my wife, daughter, and me.)

So, off I went to a new role in the Department of Public Works office, working under the senior civil engineer with several other engineers and draftsmen. While I was sure I would have a role in the big, new projects, I again learned that with no design experience to show, I would get no design tasks. This, then, was a nice opportunity to sharpen my less-than-artistic drafting talent on elements of work that were unrelated to design. Even worse, I was aghast to learn that the big design efforts were being awarded to consulting engineering firms and weren't even done by the city departments.

So I learned where I thought I should apply to become the design demon I wished to be.

At my next job, I was the newest, least senior engineer at a consulting engineering firm. It was a civil engineering office in the San Francisco Bay area. The firm had good projects that had been awarded to them by cities, counties, and universities. I had real design assignments for real projects. So began my consulting engineering career—sort of.

Once again, it became clear that my education did not equip me with the skills that could be gained only from experience. As the lowest rung on the firm's employee ladder, I ran blueprints, completed errands, did minor surveying as a rear chainman, and did some drafting of designs created by the experienced engineers.

The two things I learned at this civil engineering firm were, one, consulting civil engineers solve real design problems, and two, civil engineering firms may or may not do structural engineering design.

Achieving Zero

Complex structural design work was performed by consulting structural engineers. My hopes were to be involved with solving real structural problems, and I was sad to find out that structures were not on the firm's agenda. Oh well, more learning.

Still believing I could be a good structures designer, I took my lessons from the civil design firm and a little more experience, and sought a consulting engineering firm that designed structures. My fourth job was again at a consulting firm, as I was convinced that was where real problems were solved by real engineering firms. This firm did both civil  and structures design and had three engineers as partners, with two of them being licensed structural engineers. While the firm was small, it did big projects. The benefit of working at a small firm became very clear, as I was involved in some way in almost all the projects.

I now was also eligible by age and experience to take the professional exam to become a licensed and registered civil engineer in California. I passed the exam on my second try. I could now design things and stamp my license number on the drawings.

Fate stepped in at this firm to change my career. The only nonstructural engineer became a strong mentor for me, but he become ill with cancer. As he fought cancer, I became more competent and experienced in civil work. While I did design two roadway bridges and some timber structures, my gains in civil experience were much more extensive. On the death of my mentor from cancer, I took on some civil leadership under the structural partners. This death in the partnership changed the working environment for me. It then became my desire to move on to another consulting firm.

Achieving Zero

In 1966, I joined a large, multidisciplined consulting firm. Now I was exposed to planning, landscape architecture, architecture, and surveying, along with civil, structural, mechanical, and electrical engineering. An engineer I had worked with at my prior firm and whom I had collaborated with on a project became my boss, friend, and again my mentor.

As a senior vice president, he was instrumental in my growth in project management and the knowledge of the business of consulting, and he was an advocate of my skills. When he relocated to establish an office in Seattle, he soon asked me to come to Portland and do the same, to help the firm establish a presence in the Northwest region.

So, with a home recently purchased in the Bay Area, an eight-year-old daughter and sons three and four years old, we leased out the Bay Area home, rented one old house at first, and then bought another in Lake Oswego, eight miles south of

Portland. We became residents of the Northwest, and from 1969 to 2011, Lake Oswego was our home.

Oregon was a cultural and environmental shock to us Californians. Oregon's governor was telling Californians to please visit but not stay, and we did not feel very welcome. To add to that, Oregon had experienced the worst, snowiest winter in twenty years, followed by a cold, damp, dark spring—and summer. While we were trying to adapt to Oregon, we complained about the lack of sunshine.

We did have a brief project-related relocation to Hawaii. One late October day in 1969, my mentor called from Seattle to tell me that the president of the firm and the director of operations wanted to come and visit with me. It seemed a big project had been given to the firm: we were teaming with a Honolulu firm to design the infrastructure of a major resort community. The firm needed a project engineer to relocate to Oahu. Right now!

I called home, and while I was hearing the normal complaint of "where has the sun gone?" I asked my wife if she would like to move to a warmer climate. On explaining that the president and DOO were on their way to our newly purchased home, where we would hear about their proposal, she settled down and began to think warmer thoughts. So, once again, the entrepreneurial spirit championed our direction, and we left our newly purchased Lake Oswego home and leased a home on Oahu.

Of course, we still had our California home. Having three house payments that totaled more than my annual salary was a little daunting, but the company would pay for the Hawaii home. They also rented the Lake Oswego home to the engineer who replaced me in my job in Lake Oswego. The three-home financial dilemma that came about during our Hawaii adventure was certainly interesting, but it belongs in a different book.

The proposed eighteen-month project assignment was completed by the design group of our two firms in less than a year, and with significant financial success. We then returned to Portland and our Lake Oswego home.

One of the various infrastructure elements designed by our team for this resort project was a water reservoir or tank to be situated on a ridge top above the home and hotel sites. Having viewed many such water tanks sitting prominently on ridge tops, I persuaded the owner to move the tank farther up the ridge, but at the same elevation so it could be cut into the ridge and be less visible and not an eyesore. My proposal was accepted and the design was completed. With all the team's design work finished, I left Hawaii before construction began.

On a vacation trip to Oahu several years later, I visited the development site, and much to my enjoyment, the water tank had been built as we designed it and was nowhere in sight. When designs work well, you must enjoy the fruit of your labor.

Though the firm was very supportive and appreciative of my civil engineering skills and my mentor had put me in charge of the Portland area office, fate again cruelly stepped in; my mentor left the firm due to health problems. This gave me more responsibilities, as I was placed in charge of both the Seattle and the Portland offices as the regional manager. I became a vice president of the firm, a stockholder, and a member of the board. More and more on-the-job training.

The Portland office was a reasonable financial success and had grown to about forty employees by 1976. The Northwest region was profitable, with the profits going to the California headquarters. However, the California offices were not making a profit, so the stock I was buying in the firm kept decreasing in value. It seemed to me that something should change, and it became clear to me that it was *me* who should change.

Chapter 3
A firm is Born

"While the sorcerer slept the apprentice decided to play."
– Alan Parsons Project

After ten years with the major firm, I began a new direction with little more than high hopes, a supportive draftsman-designer, and a loan from my father-in-law. David Evans and Associates, Inc. (DEA) came into being. My designer-draftsman, David Gould, later documented this momentous event creatively by writing a "historie":

Achieving Zero

Once upon a time, long, long ago, a bright but aggressive engineer-knight and his draftsman-servant suddenly realized that they were miserable. True, they both were comfortably cared for by a large engineering-kingdom, but the king was in a far-off land (California) and they were both slaving hard in a wilderness outpost (Portland, Oregon) and had to send much tribute back to the king.

The bright engineer had many good ideas about how he could increase the annual tribute, but the king was afraid a coup might occur if the bright engineer succeeded. So the king kept the bright engineer distracted from his ideas with endless requests for more good deeds with fewer knights until the bright engineer and his draftsman-servant came to the aforementioned state of misery.

After consulting with the local wizard-developers, the bright engineer and his draftsman-servant were convinced that the omens were favorable for their establishing a kingdom of their own in the wilderness. During the Great April Fool's Day Cel-

ebration of the year 1976, while all the kingdom was reveling, they slipped unnoticed from the outpost and set forth to win a kingdom of their own.

The favorable omens on that April day in 1976 included the national housing boom being in full swing and the Northwest region providing lumber and plywood for the whole country. It also included a very aggressive financing program from companies called savings and loans.

Filling the housing booming needs required several engineering design and construction efforts. Streets, water systems, sewer systems, drainage facilities, and even playgrounds and parks were all necessary to make a housing development come to life. We civil engineers were in demand

and rubbing our hands together in enjoyment of this exciting boom. Designing land development projects and creating lots for builders to put houses on required the skills Dave Gould and I had been honing for the last eight years. We were more than ready, and we were very skilled at this part of the engineering profession.

It also helped that the demand for these professional skills far exceeded the capacity of the design firms in the Northwest. If you had a hammer and an option on a piece of vacant land that was zoned for housing, savings and loan officers deemed you a "developer" and surfeited you with funds to create lots. These well-funded, newly created entrepreneurs lined up in engineering offices to get help in moving their projects through the approval process of the local agencies, to have the plans created for the necessary infrastructure (streets, water, sewer, drainage, landscaping, and so on), and then to proceed to construction. The construction phase also required civil work: construction management, inspection, surveying, and mapping.

In our efforts to have the newly established entity of David Evans and Associates, Inc. (DEA) perform as much of the work as possible for the eager developers, Dave Gould and I began our first work days at 5:00 a.m. and ended them at 8:00 p.m. And we did it every day, Saturdays and Sundays included, until it was clear to both of us that maybe we could afford to hire some help. Our schedule could then be revised to a more normal five- to six-day week of maybe nine or ten hours a day.

With the housing boom continuing in 1977 and '78, finding help was a new challenge. Of course, one place to seek some help was from our prior firm. While it was busy too, some of the staff felt our little entity might be a better star to hitch their wagon to. So the consulting engineering firm of DEA began to grow. The savings and loan industry was creating clients faster than the engineering community could complete their work, which created a very nice scenario for us. We moved to a larger space and hired people from almost everywhere. Some of them were very talented and some, well, we needed workers.

Oregon and Washington had very affordable housing sites in the late '70s, and our clients wanted help all in both states. When our first and largest client acquired the development program for a thousand-acre parcel in Snohomish County, Washington, they asked if this was perhaps too large a project for our little firm. Well, we said, "Bring it on!" A key engineer, who had joined us earlier, agreed to be our leader in the Puget Sound region of Washington and establish our Bellevue office.

As we became the trusted adviser to this same client, he asked us to tackle a huge project in Eastern Washington, where the Washington Public Power Supply entity was constructing three nuclear power plants. Thousands of ironworkers and construction people would be building these facilities for ten or fifteen years. Each of them wanted a new home, on a new lot, in a new subdivision. So we asked another of our new key people to move to Eastern Washington to create our Kennewick office. Of course, the savings and loan industry was happy to finance these giant endeavors.

We were now a geographically diverse firm only two years after we started from zero. We had business offices in three places and about seventy people busily employed. All our work was for developers of subdivisions for housing, and all that work was financed by savings and loan institutions.

A brief return to our "historie" by Dave Gould seems appropriate here.

> *Our beginnings were almost strictly land development-oriented ... lots and lots of lots. Dave Evans designed them and calculated them on an HP-23 calculator, our first major acquisition of business equipment. He and I then drafted them.*
>
> *Even then we seemed destined to perform the extraordinary. One of our first projects involved*

measuring the surface area of the spots on a horse in order to have him certified as a Pinto. This required getting the horse to stand still (try that with your horse!) while I traced the size of the spot onto a sheet of vellum. Back in the office, the area of the spot was determined by a planimeter, a standard graphics tool, and the results certified and provided to our client. Voila! A horse officially becomes a Pinto!

But I digress. During the years 1976 to 1979 (the Golden Years of land development), we completed nearly seventy-five subdivisions, a total of nearly 4,100 lots! This works out to be an average of about nineteen subdivisions and 1,025 lots per year. In order for the two of us to do all this work by ourselves, we did not sleep between April 2, 1976 and sometime around November 15, 1979. Actually, as you can tell from the numbers, this was a period of incredible growth for our company.

We worked on subdivisions from A (Alpine Ridge) to Z (Zoon Subdivision). We did a little bitty, micro-subdivision (Flemming Terrace, four lots), and great, big, mega-subdivisions: Kennewick Park and Canyon Lakes (with thousands of lots).

We worked on developments that were set on hillsides so steep they said it couldn't be done. After completion, I think they were right, but the views were great.

We were directly involved in all the large community developments in the Northwest. Yes, these were the Golden Years; life was good. The land developers were elbowing each other for position in our lobby, we had many new employees, and Jimmy Carter was in office. In short, something bad was sure to happen.

Chapter 4
Bumps in the Road

"Let me tell you a story. The Devil he has a plan."
– Talking Heads

The tale of the plight of a consulting engineer would be less than useful if it did not include hard lessons learned. In short, something bad did happen. Really bad.

In 1981, the savings and loan industry imploded. No more money for budding developer-entrepreneurs. No more loans for homebuilders. Interest rates went from single digits to near 20 percent, and inflation was rampant. Clients just disappeared and didn't bother to pay their bills.

Achieving Zero

This was not my idea of how the business world worked. My naïve belief that having three offices spread across the Northwest would provide the geographic diversity needed to absorb market fluctuations and stay afloat was just that—naïve. Not only did the savings and loan funding of the development boom end, the Washington Public Power Supply (WPPS) system also came to an abrupt halt. WPPS became WOOPS. The Eastern Washington economy now survived on the unemployment checks of the thousands of laid-off construction people.

During the four Golden Years from '76 to '80, I hired landscape architects, surveyors, and planners, along with civil engineers and draftspersons. I had about seventy employees, with enough continuing work for maybe fifteen or twenty of them.

I had not been out of a job, terminated, fired, or laid off since I was first employed at PG&E. Now it was my task to meet with each of the people we could no longer employ and tell them that their wagon was being unhitched from our not-so-bright star. And I needed to have this talk with each of the fifty employees we had to lay off. This was one learning experience I have never forgotten and one I never wanted to repeat.

To help the firm survive, we regrouped around twenty workers and paid ourselves only for our productive efforts. I went without pay, and reduced hours and reduced pay was the norm for all. I think you can say this had a dramatic effect on our culture.

In 1982, we were about a twenty-person firm, still in three locations and surviving on minor programs that remained to be finished in land development. We now had an urgent need to seek non-land development assignments from a new set of clients. Having learned that engineers do not live from developers alone, I set out to diversify our client type and diversify our services. Our engineers, planners, landscape architects, and surveyors were skilled and certainly could do nondevelopment work. We just needed to find some.

It was also clear that the Northwest was recovering at a much slower rate than the Sunbelt states of California, Arizona, and Texas. Our diversification within Oregon and Washington was not helping the firm's survival at all. If I wished to sustain our core group and provide work so they would stay employed with us, expansion outside the Northwest was required, providing a new challenge for this young firm.

In hindsight, it is clear that when the times are going well, you need to ask, "What if this golden era ends?" Some strategic marketing planning may have made the bumps in our journey less painful, but that was not yet one of my skills.

Chapter 5
Some Surprises are Good

"If you could read my mind, love,
what a tale my thoughts would tell."
– Gordon Lightfoot

I began searching and traveling throughout the Sunbelt, hoping to find an encouraging opportunity for a new DEA office location. After chasing many dead ends and becoming frustrated, I came across a small civil engineering consulting firm in Southern California on the verge of closing. The eighty-year-old sole proprietor had been slowly downsizing the firm to about twelve employees, but it had clients and work. And *what* clients! To this consulting engineer, who had spent the last five or six years working only for developers, the clients—Pacific Bell, AT&T, and California Edison—were golden. So I made a proposal for acquiring the firm and employees, we negotiated, and a deal was completed.

Acquisitions require time, energy, and significant due diligence, and due diligence turned out to be a significant challenge in acquiring this Southern California team. The sole proprietor requested that we buy his building and acquire all his related assets in an all-cash deal. For DEA, just coming through the recession/depression in the Northwest region, this was not going to happen.

After several meetings and lunches, almost always away from his office and out of earshot of his staff, I proposed a plan to hire all his people, rent his building, buy only the furniture and equipment needed to stay in business, and leave him with the remaining significant assets. These were mostly accounts receivable from the three big utilities (good as gold!) and a sizable amount of cash on hand. We would also pay him, for those three utility customer accounts, a small percentage of revenues from future work performed by the newly acquired workforce. This was not the tax-friendly plan he wished for, but it let him divest himself of responsibilities of operating a diminishing firm. We signed an agreement, and with that business behind us, he allowed us to speak to his twelve employees.

So our CFO and I went to the office to share the great news that the employees would now be part of a wonderful firm headquartered in Oregon. What had seemed like a great idea turned into a great shock when the office lead engineer quickly informed us that all twelve employees were well aware that the sole proprietor had been planning to sell or close the office, and all twelve had made contingency plans in the event of that sale or closure. And *all* were quitting.

This was not my beautiful plan. That Monday morning Welcome meeting was, according to their agenda, to become the two-week notice that all the staff would be gone. After surviving the initial shock and explaining a little bit about the wonderful company DEA was, I asked them all to just give us a chance. Luckily for me, these people were excellent workers, strong technically, experienced, but not entrepreneurial. All of the entrepreneurial types had left the firm long before. They agreed to give us the opportunity to convince them we were as good a contingency plan as picking up and quitting.

As the CFO and I concluded our visit and had said as many encouraging words we could come up with, we headed back to the airport, shaking our heads and shaking in our boots. Was this an acquisition or a disaster? The good news turned out to be that they all stayed; several stayed all the way to retirement and some are still with the firm.

My forecasting and planning to establish a Southern California presence with this team quickly evolved into a fiber-optic SWAT team that would move to the location of the many fiber-optic cable projects that were starting up all over the western region. The team moved to Washington, to Oregon, and back to California to design and manage major long-line fiber programs. As an early adapter and a knowledgeable group, we were very, very successful in the fiber-optic cable frenzy to complete a network. As Yogi Berra said, when we "came to this fork in the road [our plan for a Southern California office], we took it." Taking that route helped DEA become a national and international designer for fiber-optic facilities.

We were now a firm in three states and expanding our experience into the utilities market. In 1985, the growth of telecommunication facilities via fiber-optic cable was just beginning. We were once again growing, thanks to the need for civil engineering services to get fiber-optic cable in the ground as fast as possible—or even faster. Even if you have excellent skills and a good plan, luck is very, very nice. Our little Southern California team, skilled in telecommunications design, would lead us into our next golden years.

Chapter 6
Defining the Firm

"If you build it, they will come."
– From the movie Field of Dreams

As we enjoyed and grew with the booming telecom market that required all of the civil engineers' various talents—design, survey, construction management, and so on, plus environmental analysis and land planning—a major event in our firm's culture occurred. In an effort to expand our planning capability, we advertised for senior planning people. After reviewing numerous resumes and several applicants, we narrowed the choice down to two talented people.

I believed either would be a great addition of a talented professional to our team. While our workload required only one new planner, making the decision of which one to choose weighed heavily on my mind. In the first dawning of what would become our credo and eventually our formally stated philosophy, it came to me that if these two planners were truly the outstanding professionals that I envisioned them to be, we should hire *both* of them.

If we could give each of them support, encouragement, and a limited amount of freedom, they would make their own success under the firm's umbrella—and their success would be success for the firm and our clients. And so we did it. Together they completed the work we had on the books and then proceeded to find new work to expand and grow. What seemed novel at the time—to have more people than you need—became the instrument of our growth and success. Today all our promotional materials, our website, and all the documents about the firm begin with our philosophy statement:

We find outstanding professionals and give them freedom and support to do what they do best.

I have always told our employees at meetings where the only real purpose was to let people ask me about the workings of the firm—commonly called Stump the Boss meetings—that each of our employees are not working for the firm, not working for me, but working for themselves. Their efforts, their hours on the job, and their intelligent solutions to problems were all solely for themselves. By their work they could provide for their lifestyle, their families, and their future. The firm supplied only the framework and opportunity for them.

Of course, the firm benefits from their diligent efforts, but the reason for the effort must be so that each could achieve the secure, healthy life that they deserved. Believing that employees are working for themselves makes the challenges and accomplishments they achieve more real, more enjoyable, and less tedious.

The business of consulting engineering is a people-friendly business. The efforts of employees are recorded in time units. Workweeks of forty hours are the norm, but longer hours are often required to meet deadlines. And while the work is time intensive and time sensitive, it is also very flexible.

From the first days, our firm has allowed workday flexibility for people. I often recited that I wanted no employee to miss an important date or time in his or her personal life due to the demands of the job. I tried to never miss my daughter's or sons' athletic events or academic presentations, and I did not want any of our employees to miss the important events of their real life. Opportunities to complete any work or business effort that was placed on hold while an event was attended is and was always available. Hey, what are you doing between two and three in the morning? This flexibility and family-friendliness of the consulting business is just one of the things I believe make it very special and unique.

With our philosophy of hiring outstanding professionals whether we had a need or a job for them or not and telling them they are working for themselves and that this is a people-friendly business, well, our recruiting worked reasonably well. Recruiting also required that we tell prospective employees about the things the firm believed to be significant and important.

My struggle to define a strategic plan and direction focused on the culture and values employees would be exposed to and, I hoped, ones they would embrace. Instead of a plan, I institutionalized the five values that would define our firm. As these five values became known, shared, and posted on office walls and in our promotional materials, two more values emerged and were added, because they too were part of who we were and who we endeavored to be.

The values are the subject of chapter eight, and defining the culture and values and philosophy helped to define the expectations of new employees. If you liked our makeup, this could be a good place to hitch your wagon and ride it to your own future.

As the description of "outstanding professionals" came to be extremely broad, our technical capabilities grew from civil, infrastructure design engineers, landscape architects, and land surveyors to architects, planners, structural engineers, traffic and transportation engineers and planners, environmental specialists—botanists, biologists, ecologists (better known as the "bugs and bunnies" team)—and hydrographic surveyors.

The hydrographers provide a great topic to expand upon. With Portland being dependent on barge and ship traffic on the Columbia River and also being the home of a Corps of Engineers district, there was continuous work available in hydrographic surveying. Surveyors helped to ensure that ongoing dredging projects met the needs of the many types of ships using the river.

David Evans and Associates therefore became boat owners. The boats became the platform for incredibly expensive hydrographic toys that increased our capability to map, search, calibrate, and analyze the bottom of a waterway. In addition to monitoring dredging work, we located anchors lost from container ships, mapped cableways for underwater cables, and mapped proposed outfalls to the bays and coasts of Oregon, Washington, and Alaska.

This team's reputation as experts grew to the point it was asked to gather its equipment and travel to the Philippine Sea to aid explorers searching for a lost Spanish Galleon—filled with gold! While I would like to report an actual finding and that we received a share of the gold, well, we discovered a trail of debris that we believed held promise, gave the results to our treasure hunting client, and returned from the adventure never to hear more about it.

Not all underwater searches were fun. When two F-16 Air Force fighter planes collided over the Columbia River in July 2004, we were requested to come to the scene and assist in the recovery. And we did.

Our growing diversity in people and services added greatly to our tale of consulting engineering, and it added depth and security for all the various design engineering professions that make up the firm of which we are a part.

Chapter 7
Stepping Into the ComPuter Age

"I've done some bad, I've done some good,
I've done a whole lot better than they thought I
would, so come on and treat me as you should."
– Don McLean

Now it seems right to insert more of Dave Gould's "historie," the continuing chronicles of DEA.

Our beginnings were rooted in the feverish days of the land development boom. Life was good; we would work night and day producing construction drawings and plats, and the supply of developers seemed endless.

Achieving Zero

This frantic pace occasionally resulted in small problems. During the Grand Opening of the initial showpiece phase of Parkview in Somerset West, it was rather rudely brought to our attention that there did not seem to be any water mains in place. Sure enough, not a hydrant in sight. The whole affair just ruined a perfectly good ribbon-cutting ceremony. We clearly needed help.

We grew from two people in 1976 to twenty people in 1977, forty people in 1978, and seventy-five people in 1979. Then, one day we noticed...there were no more developers in our lobby. Our staff plummeted in 1980 and fell even lower in 1982. Some of the larger public works projects we had been awarded were placed on hold, interest rates and inflation were soaring, and property values were dropping.

Diversification saw us branch out into various adventures: acquisition of a surveying company and a Southern California engineering firm.

The years 1982 to 1986, the beginnings of the Reagan Administration (no, I'm a Democrat), saw our little firm grow from the aforementioned twenty-five people to sixty people. At the end of this period, less than half of our work was subdivision related. We were marketing ourselves like the dickens to more stable, less economy-driven clients—cities, counties, and other public agencies—which led us to larger, better-financed private projects. As a result, today any really, really big project of nearly any kind happening in the Northwest usually has us involved in some capacity.

This part is going to focus on computers and their contribution to our history and our future. I have seen the movie 2001—A Space Odyssey (an IBM training film about what can happen if you don't properly care for your computer), and I can generally tell the difference between a computer and, say, a toaster. My credentials thus justified, I will digress for a moment to an earlier time and the early generation of computers used in the engineering and surveying industry.

My first exposure to the computer was the LGP 30 (Librascope General Precision) series utilizing

Achieving Zero

the SASSY software (Subdivision And Surveying SYstems). The LGP was about the size of a chest-type freezer and about as attractive. These machines were large, noisy, and slow, and more importantly, they had no little colored blinking lights, which are so essential to today's computers.

When Dave Evans started DEA, he made a commitment to stay on the leading edge of technology: Our first order of business, as discussed earlier, was to purchase our first computer, the HP-25, a hand-held model with "memory." This had little teensy keys and a little teensy display and could be programmed manually. We used it to do all of the computations for our first subdivision. (We both now need corrective eyewear to read.)

We next acquired a series of Hewlett Packard's personal computers. These had the advantages of easy-loading software—either a tape or a disk—that had to be loaded into RAM each time before you could use it. We also noticed that by altering the software slightly, we could tailor the output to suit

David F. Evans

specific needs rather than having to sort through the "general" output for answers. This was to have great impact on our future computer operation and lead to a "discipline" that is truly unique in our business.

While all of the computers we had purchased and used to that date were satisfactory for engineering and surveying, all of our accounting was still being done by hand—and we were growing. Dave's quest for leading-edge technology and little colored blinking lights resulted in the purchase of an IBM Series 1, a computer so advanced it had to look back to see the leading edge. It had only one minor problem—no software.

When it arrived, Dave lined us all up and asked for volunteers to take one step forward. All of us more experienced personnel took one step backward, which revealed our very first computer programmers—two rather slow-footed, new, entry-level engineers who are no longer with us.

They were the core of our programming discipline and got us hooked on creating custom software that would make computers do exactly what we wanted, without the burdensome and time-consuming output of superfluous data. They would spend all day programming like crazy, and our Series 1 would blink its lights happily. And all was well.

This period (1980 to 1987) was noteworthy in that the personal computer was in its infancy, and there was very little good engineering software available. For us to make efficient use of our computers, we were forced to write our own stuff.

At this point in 1987, we had between 300 and 350 people (the number seemed to fluctuate daily, even hourly). We had experienced a 50 percent increase in population with a 200 percent increase in computer stations within the same period of time. The software industry had caught up with us, and our computer team was hard put to constantly

manage station locations, make repairs, and order new computers for the firm at large.

A business that began with four sawhorses and two doors that served as our drafting tables had become a thoroughly technology-driven enterprise. The computer age had increased the speed of all phases of the profession and eliminated at least one: draftspersons were, like dinosaurs, suddenly no longer with us. While the culture of consulting was still user-friendly, technology had made it less personal, thus reinforcing for us that a strong firm culture was essential for continuing success.

Chapter 8
Clarity of Purpose

"Brave Helios wake up your steeds, bring
the warmth the countryside needs."
– The Moody Blues

David Evans and Associates has an ideology. It has grown over the years from the philosophy and the first five values to seven values, to include a core purpose, a vision, and a vivid description of achieving that vision. A firm that envisions itself as still existing in some form in the year 3000 (Y3K) must have a good understanding of what it is all about and what it pledges to become.

As I stated in chapter five, we began to institutionalize our values while trying to define a strategic plan. It seemed to me that before we could chart our course forward, we needed to know what things were strategically important to us. So the first five were cast in stone in 1991, with clear descriptions of their meaning.

1. Honesty
   We must be scrupulously truthful with our clients, our coworkers, and ourselves in our professional and personal actions and work.
2. Consideration
   We must be concerned for each other and act with respect and sensitivity toward everyone with whom we interact.
3. Openness
   We are unafraid of sharing good and bad news and believe knowledge of all aspects of our company, if shared, will strengthen our efforts.

4. Enjoyment
   We recognize DEA is not the center of the universe. Enjoying our efforts and our workplace are essential to maintaining our professional excellence.
5. Involvement
   We are good citizens. We volunteer for and provide professional advice to a variety of community projects and enterprises.

As we asked all our employees to know these five and strive to live them, two more values came to the forefront as already somewhat embedded in our culture.

6. Entrepreneurial Spirit
   We are keenly interested in new opportunities. We venture into new fields and develop new approaches to our work. We do not take these risks lightly, but seek input from our colleagues and clients so that all who are af-

fected are aware of the risks before action is taken.
7. Financial Security
   We pride ourselves on providing for the well-being of DEA employees and their families. To do so, we prepare adequate budgets and work within them in order to produce the profit needed to attain financial security.

From these sprang a need to define why we should be in business at all. Are we about making money? Are we about providing excellent design services? Are we important to society? To humanity? Through a collaborative effort, a reason for DEA to exist evolved.

Our Core Purpose

To improve the quality of life while demonstrating stewardship of the built and natural environments.

That core purpose led us to define a vision and vivid description of how we would proceed in bringing that noble purpose into being:

Our Vision

To work together, doing whatever it takes, to be recognized as the best by our clients, employees, shareholders, and peers—locally, regionally, and nationally.

Our Vivid Description

We will achieve our vision:
- When all of our employees are highly satisfied with their jobs, get a great sense of accomplishment from their work, do challenging work, and are committed to DEA as a career choice.
- When our clients and peers perceive us as the best and respect us for our innovations

and creativity as well as overall client services and quality.
- When we move from an economic company to a living company where we can attract, grow, and retain knowledge and fulfill our potential.
- When we employ a new model of practice, from selling hours to creating value, as judged by our clients. In this sense, value to us means not only monetary, but includes growth to our employees, learning, leadership, a sense of community, and the development of a healthier, more sustainable environment.
- To measure these, we are using employee and client surveys and modeling our incentive compensation program about achieving specific returns and a specific financial performance. Each of the three—employee satisfaction, client/peer satisfaction, and financial performance—has equal weight in the program.

This core ideology is posted on our website, on posters in each office, on pocket cards for all employees, and, I hope, in the hearts and minds of every DEA person.

Doing the Right Thing

When I composed an ongoing series of columns for distribution within DEA, a significant portion of my internal writings were intended to keep these core values at the forefront of people's thinking. For one such essay, I was writing about honesty and came across a code of conduct composed by Archer G. Jones. After sharing it with all, I posted a copy on the corner of my desk. Everyone needs help keeping great thoughts at the forefront of their minds.

There is but one rule of conduct...to do the right thing. The cost may be dear; in money; in friends; in influence; in labor; in a prolonged and painful sacrifice. But the cost not to do right is far more dear; you pay for it with your integrity; in honor; in truth; in character. You forfeit your soul's content; for a timely gain, you barter your soul's infinity.

Being scrupulously truthful is certainly "doing the right thing."

In our endeavor to be good citizens, we have supported public television and hosted their telephone support program. We also provide direct support to local schools. We have provided cash, readers, speakers, and teacher's helpers to the elementary school closest to our office. Our value of *involvement* has been an active one. How this support for the nearest school came to be has been a topic in several of my speeches to professional societies.

In 1989, my youngest son graduated with a degree in architecture from the University of Notre Dame. Of course, my wife and I flew to South Bend for the event. The commencement speaker was Peter Ueberroth, then of Olympic fame and the future commissioner of baseball. A portion of his talk related to the need for involvement in education by everyone. He said, "You businessmen in the audience, don't you believe that you can leave the responsibility for education of our youth to the public education system. You must be involved, not just in PTA and attending school events. When you return home, visit the school nearest your office and ask how you can help."

Although this graduation of my third and last child from Notre Dame meant tuition was finally over after twenty-three person-years of higher education expenditures, I couldn't agree more with what Peter had said. Upon returning home, I got out a school district map and located Abernethy Elementary School, K-6, about two miles from our downtown office. I called the principal and asked for an appointment to discuss support for the school.

It became clear that he had no idea what I was thinking about when, at the appointment, I relayed the Peter Ueberroth story and said I was there to be the school's fairy godfather to the extent our firm could afford it. So, what would he wish for the school? He didn't quite faint, but he pinched himself several times. Thus began our relationship with the Abernethy Eagles.

Initially, the principal asked us to print business cards for his teachers. The school district didn't provide them, and because we print our own in-house, this was a snap. We then provided volunteers with their HOSTS program. HOSTS is "Help One Student To Succeed." Our volunteers provided reading support on a one-on-one basis. We allowed our workers (including me) to take the time during regular working hours to go to the school and sit with fifth- and sixth-graders and listen to them read. This wasn't paid time, but it could be compensated by staying late, coming in early, or taking a vacation hour.

We participated in this program until the school decided on a different reading effort. Everyone who spent time at Abernethy received much more in satisfaction and good feeling than the small effort involved on our part. Over the years, we also bought wooden blocks for the kindergarten, tricycles and scooters, Franklin Spellers and Weekly Readers, and we donated miscellaneous printing. Lastly, their principal asked if we would financially support the hiring of a science teacher, as the school had no science program due to funding decreases. We budgeted about two thousand to three thousand dollars per year in direct financial aid for the hiring of that teacher, along with our volunteer effort.

Our effort for Abernethy continues even today, and it makes us feel good about ourselves. But what happened beyond our own good feelings was not expected. We received thank-you notes from the kids, an honorary membership in the PTA, the Principal's Award for Volunteers, and a plaque and letter. In 1991, the school district honored our firm as Volunteer of the Year, and in 1993 we received a Most Valued Partner Award from the chamber of commerce and the school district.

These responses from the school made headlines in the Metro Section of the state's major newspaper, *The Oregonian*. I believe seeing such a positive headline about our firm raised our profile significantly in Portland. It certainly did in one person's eyes, as a letter to the editor was published in the Oregonian soon afterward. The author wrote, "To those who lament the lack of heroes in our world... there was a story about David Evans, president of a local company, who adopted a school...and taught us what heroism is all about." We had achieved the title of heroes. In print! For doing what began only as an effort to help where help was needed.

Our effort may have helped us achieve only a zero, but it also gave all in the firm a good, warm feeling.

Learning and the DEA University

If there is one specific course of action that is essential for a consulting engineering firm, it is to clearly define your purpose for being, your values that create your culture, and your vision.

So, what are we doing now to sustain our culture? We have two well-defined, firm-wide initiatives that continue to create and perpetuate our culture.

First, we have established the DEA University. Although our outside adviser said he knew of no firm of less than 1,500 people that was able to create and sustain an in-house university program, we blundered ahead. In addition to being Chairman Dave, I also became Chancellor Dave.

Our university campus is the firm itself. Just as at most universities, learning takes place in lots of settings outside the classroom. We see becoming a learning organization as the goal of the university. Our training and educational programs occur within four colleges around what we call "key activities," which are founded on our cultural elements.

Every DEA employee is enrolled, has a professional development plan based on his or her career desires, and has a curriculum that is tailored to his or her plan. The university has established a charter and has established deans—and is selling T-shirts.

The second current and ongoing initiative is based on the book *The Oz Principle: Getting Results through Individual and Organizational Accountability* by Roger Connors, Tom Smith, and Craig Hickman. We have adopted the Oz Principle and its description of accountability in our efforts to further improve and adapt our culture to fit changing times. This is not a change to our values, core purpose, vision, or philosophy but an effort to have these cultural elements become more meaningful for each of us. The Oz authors define accountability a little differently than the normal approach of "Whose fault is it?" or, as Lucy in *Peanuts* said, "When you're down and out, lift up your head and shout 'SOMEBODY'S GOING TO PAY FOR THIS!'"

The Oz theory focuses more on the question of "What else can I do to achieve the results we are seeking?" Subsequent to writing the Oz book, the author created the firm Partners in Leadership, and they are helping our continuing efforts to inspire our culture, to achieve the results we desire, and to achieve our vision.

If our culture can create the experiences people enjoy and believe in, they will hold the beliefs we say are required for success. Those beliefs will create individual actions, and we will have a culture of accountability that will provide the results we hope for.

Chapter 9
Building a Board of Directors

"A man's got to know his limitations."
– Dirty Harry Callahan, played by Clint Eastwood

By 1991, DEA had accomplished several small acquisitions. Each one was part of my continuing effort to capture outstanding professionals. These acquisitions brought us leaders of firms that wished to focus on professional practice and not on the administrative side of running a business. I am not a big fan of accounting, human resources, personnel issues, and so on, so I was pleased to offer these firms the centralized administrative support systems of DEA so the leaders could do what they do best: be outstanding professionals.

All of these acquisitions required an action by our board of directors to legitimize the deals. I sought some legal advice on crafting the acquisition agreements, but mostly they were my thoughts and plans and my understanding of the unique concerns of each acquisition target. Getting the board's approval was not an issue as the board was me, Dave Gould, and another long-term leader who had hitched up to my star years before. Voting on the acquisitions took place long after the fact, when the attorney created the state-required corporate meeting minutes.

In 1991, I proposed and consummated the largest acquisition I had ever attempted, and the size and complexity of that transaction brought home the thought that maybe I needed some business advisers and some oversight.

The firm we were acquiring was located in Arizona and had about one million dollars in accounts receivable on their balance sheet. Many were 90 to 120 days old. I believed the leader of the firm when he said the accounts receivables were all good. We completed the deal based on the value of the balance sheet, and a significant number of DEA shares were provided to the owner-leader.

It turned out that the accounts receivables were not good—they were worthless—and we wrote off the one million dollars over the next year or so. I should have had the seller guarantee the accounts receivables, but even though our attorney cautioned me, I did not. This was a sad and hurtful learning experience—and expensive. It was clear that my business knowledge needed significant help, but because DEA was a team of professionals in technical disciplines, that help would have to come from outside the firm.

With the help of a newly retained business counsel, I sought out four outside members to join our board of directors. These were four very senior people from nonconsulting firm backgrounds, who all agreed to provide the oversight I badly needed and was overjoyed to have. Bringing our board up to seven members, with four of them being non-employees or outside directors, was one of my wisest moves. I now answered to and served at the pleasure of a board of directors, with a majority of its members having no allegiance to me. Their primary concern was for the welfare and success of our whole firm and its shareholders. This shift opened up a very different and life-changing role for me, but it was wonderful for the firm.

Our board became the governance entity that operated in the same way that boards of public companies do, but we were (and still are) committed to employee ownership. The advisory nature of the board placed the four outside persons in the position of not having the power to veto actions, but being able to offer their best advice, for example, saying, "I don't think that is a good idea or plan." Their expressed concern was all that I needed to stop that program or at least to make the revisions they suggested. I cannot remember ever having to ask for a vote on a program that was under debate. If I could provide the reason for my proposals and there was still a strong concern from the board, I would drop the proposal.

Of course, the board is selected by the shareholders at the annual meeting of all shareholders. In 1991, I held the majority of the shares, so I did have some security in the outcome of any vote. Our first four outside directors included a retired bank executive (not from the firm's bank), a local successful businessman, the retired CFO of a major wood products firm, and a retired executive from the state government.

Bringing in the retired state executive proved to be a problem, as that state government promptly stopped requesting our services on their new work. While the state agencies perceived a possible conflict, I had hoped for the opposite result. I felt that the state agency would recognize that we would have a better understanding of their needs. My mistake! I then had to ask that director not to stand for election at the next shareholders' meeting. Learning your limitations is not a fun experience.

While our directors are elected for one-year terms, two of the four elected in 1991 remain on the board today. Their care, concern, loyalty, and even love for the firm are evident in their continuing willingness to serve. With their twenty years of knowledge of our travels and history, they are an incredible resource to the generation of leaders now taking on the management of DEA.

Chapter 10
An Article Published by Civil Engineering News, November 1996

"Then they showed me a world where I could be so dependable, clinical, intellectual, cynical."
– Supertramp

When I was invited to write an essay for *Civil Engineering News* about the state of our civil engineering profession, I was also charged with not complaining about the "sorry state of affairs for civil engineers" unless I could offer "some constructive solutions for improvement." Well, how can you improve on what Herbert Hoover said about engineering in his "Addresses upon the American Road"

> *His is a profession where he has the fascination of watching a figment of the imagination emerge through the aid of the sciences to a plan on paper. Then it moves to realization in stone, or metal, or energy. Then it brings jobs to men. Then it adds more and better homes. Thus it spreads progress and opportunity over the land. That is the engineer's high privilege among professions.*

However, in an article in *Engineer's Week* in 1954, he also added,

> To the engineer falls the job of clothing the bare bones of science with life, comfort, and hope. No doubt as the years go by, people forget which engineer did it, even if they ever knew. Or some politician puts his name on it. Or they credit it to some

promoter who used other people's money. But the engineer himself looks back at the unending stream of goodness which flows from his successes with satisfactions that few professionals may know. And the verdict of his fellow professional is all the accolade he wants.[1]

I think we need a little more than only the accolade of our fellow civil engineers. I guess I want a lot more, a whole lot more.

I find myself in the position of Fraser S. Keith, secretary of the Canadian Society of Civil Engineers, who was quoted in *Engineering News-Record* saying,

> *Let me ask you, what position would the government of Canada be in to-day in carrying out the nation's work without the services of the men in the engineering profession. The majority of the departments of the government would be unable to operate without our help. Do the political members of the*

> government realize that this is true? It is certain that they do not. Why? Because we of the engineering profession have in the past taken no corporate action to insure that they should. Instead we have to some extent acted like dumb driven creatures, accepting the crumbs that have fallen by the wayside, content to sell superior qualities of mind and training for a mess of pottage.[2]

I'd like to ask Mr. Keith about this, but since this was published in the May 22, 1918, issue, he may not be available.

I was also told by the *Civil Engineering News* that I could do a story or parable, and that set me to thinking on how to implement solutions. It would be easier if I had authority, if I could decide, if I were king. Well, if I were king...

First, I probably could pay off all my credit cards—including my wife's (I think).

Second, I think I would make my birthday a national—I mean a kingdom—holiday.

Then I guess I would start thinking about how to improve my kingdom. It is showing signs of wear and tear. It seems we have increased our population some thirty-nine million subjects in the last fifteen years, and our 265 million people are all using our resources, our water, our roads, our airports—basically our infrastructure—and, sad but true, with 265 million users every day, our infrastructure needs significant help.

Since all the departments of my kingly staff—you know, the ministry of defense, the ministry of commerce, the ministry of the treasury, and so on—are all dependent on the ministry of infrastructure to get their work done, my minister of infrastructure would be a really busy person. He would have to be very smart, well educated, experienced, knowledgeable, and creative.

I would reward my minister of infrastructure with the full monetary resources of my country's treasury. Well, maybe I would be my own minister of infrastructure. Why not? I am a civil engineer—

and I have several licenses to prove it. I can barely visualize the joy of assuming the challenge of solving all infrastructure problems of my country, especially with the full resources that are required being available.

How would I start? I think I would first find the additional leaders and managers who had expertise in infrastructure design, operation, and maintenance—you know, other civil engineers—and make them deputy ministers. I'm already having so much fun that I'd better appoint someone to do some of my other kingly stuff, because my ministry of infrastructure probably needs full-time attention.

Well now, this is really a dilemma! I really don't think I want an attorney as "vice king" or an accountant. The doctors are not right for this. I know what we really need—while we solve the problems of clean water and air, safe highways and airways, efficient sanitary and solid waste handling, and cleaning up of prior environmental disasters—is someone to tell the world that we are doing it. The vice king will be a public relations expert. Well, vice king is probably not the right title. Maybe public relations specialist (PRS).

My PRS will know how to tell the country of the miracles my ministry will be performing. You cannot expect our subjects to appreciate the things they desperately need when they use them every day and they are almost always in working order—unless we tell them. And if we are going to expand, rebuild, rehabilitate any of our infrastructure, we want the subjects to understand not only why but how our work is essential to the kingdom's immediate and long-term well-being. Does this sound like a big job for a PRS?

Well, maybe, in addition to my PR expert, I would tell my legion of civil engineers in my ministry that they could each get a PR person too. Each one of these infrastructure deputy ministers could then be sure his part of building the fabric of our country was appreciated by all who benefited. I know each infrastructure deputy would need a PR helper, because infrastructure engineers really get too involved in problem solving to tell everyone about all the fun they are having. I don't think it is because they don't want anyone other than other civil engineers to know; it is just that the problems do take intense and focused effort and are complex and time consuming. Even as the problems are solved and new facilities are built or old ones rehabilitated, the facilities need civil engineers to operate and manage them. This just does not leave time to tell anybody how important civil engineers are.

Would you believe most subjects don't even know what we engineers do? Well, my subjects would learn. We, my PRS expert and I, would turn everybody's water off for one hour a week and say, "It is deputy minister for water works appreciation time!" Everyone would cheer when the deputy minister turned the water back on. We would be sure that, while the water was off, the deputy minister for sanitation was also being appreciated for receiving, disposing of, purifying, and recycling this critical resource. If you made the mistake of using your waste system during that one hour of appreciation time, I am sure you would cheer when the service was restored.

We would probably need to close at least one bridge every week, so people would know that bridges do not just stay up by themselves. Can you imagine how many people would become intimately familiar with civil engineers if we closed the Golden Gate Bridge or the George Washington Bridge once in a while for "bridge engineer appreciation day"?

This would probably lead to having a "road toll day" every month as well. You could not leave your driveway until you called your credit card number into the infrastructure ministry and talked to (or at least left a voice-mail message for) the deputy minister for transportation and said how pleased you were that roads exist and thanked the civil engineer by crediting the treasury account a nominal amount.

The PR deputy would put the proper spin on this so it seemed the appropriate thing, since our transportation system is essential for...everything. We would be sure that any credits that came to the treasury for roads would only be used for road and highway improvements. It just seems right.

Because it seems that every airport in my kingdom is already under construction, we probably wouldn't need a "no plane" day. The civil engineers building our new terminals and runways do need a better public relations person. I don't think that the passengers are cheering on the engineers who are trying to keep up with their dramatically increasing number. Of course, if any plane falls from the sky—heaven forbid!—the design engineers get lots of attention. My PRS expert would make sure that the attention after the disaster focused on how safe our systems truly are.

When all the subjects realized that every breath, drink, ride, and so on was dependent on the ministry of infrastructure, we would find civil engineers not just revered, but feared. If you said a bad thing about engineers, like "engineers are bad at communication," someone may turn your water off. If you complained about highway maintenance, your credit card contribution to the ministry would double.

I think I'm creating a power trip for civil engineers. Well, that's not all bad.

I bet my PR specialists will develop some catchy phrases to help people know what we civil engineers really do. I can see them now: "Our job is Job One," "This Job's for You," "We Bring Good Things to You," "Solutions for a Small Kingdom," "Engineers Just Do." The possibilities are mind boggling.

Well, I'm not going to be king, so what do I do? When our infrastructure is in great need, when our society is in the mode of "If it ain't broke, don't fix it," when I know that the only time you can fix much of our infrastructure is before its broken, I think I need to be my own public relations person.

Part of my professional responsibility is to advance my profession. I must not only have technical skills to solve complex problems and management skills so I can run the things I make run, but I also must have communication skills that reach out beyond my circle of fellow engineers. I can't delegate the responsibility to my associations or societies, for it is my profession and I must do my part.

My firm will sponsor public television and support our local elementary and high schools. We will provide scholarships at our local colleges and universities. And we will tell the local media that engineers are doing these things. We will volunteer as media experts so when disasters do strike, we can help citizens know what engineers are doing to aid in the repair or rebuilding. We will put signs on our job sites and on our vehicles. We will have a website, be part of national alliances of design firms, and write letters and articles—even ones for periodicals that aren't for engineers. Maybe even write a book!

If this sounds somewhat like old ideas and solutions, it is—with good reason. The July 19, 1919, issue of *Engineering News-Record* includes an article titled "Why Engineers are not Appreciated" that states,

> *Seven reasons for the lack of appreciation of the engineering profession by the public were given by W.W.K. Sparrow, corporate chief engineer of the Chicago, Milwaukee & St. Paul Ry. [Railway] in a recent installation address before the Cincinnati Chapter of the American Association of Engineers. They are as follows: Inadequate compensation; lack of organization; lack of initiative in public affairs; lack of publicity concerning the engineer and his worth and his relation to the better health, comfort and welfare of the community; lack of proper public regulation in the form of licensing;*

improper application of the definition of the term "engineer" and lack of general education and inability to speak well.

And from Fraser Keith again,

We have it in our hands to bring about a different state of affairs, but action is required, not words…The government and railway officials and the general public have only a vague conception of what they owe to the engineering profession for their material welfare, and they will continue in ignorance until we have educated them. In the meantime, gentlemen, it is we who are culpable, not they.

It seems that the only thing that has changed in eighty years is we now have women engineers in addition to our "gentlemen." Fraser Keith also had a solution. It was a public relations program.

Chapter 11
Understanding Our Plight

"The pattern of my life and the puzzle that is me."
– Simon and Garfunkel

Our attorney went to college. After graduating, he went to law school. Upon completion, he passed the bar exam and became "licensed" to practice law. He is now the senior principal of a large, local firm after some thirty years of experience. He charges us 450 dollars an hour for his service.

The accountant for our firm also went to college and pursued an advanced degree, an MBA. He eventually became certified as a CPA and is now a partner in a Big Four, Five, or Six firm (who can keep count?). He charges our firm 350 dollars an hour.

We hire these people because we believe they have the best skills for the tasks we need completed. We pay their fees and believe we receive good value for our dollars. In fact, I value their knowledge of our needs, their personal interest in our firm, and the relationship we have with them.

I am a college graduate. I passed the Fundamentals in Engineering exam to become an EIT (engineering in training). After four years of training and experience, I passed a license exam to be registered as a professional engineer and surveyor. Several decades later, I became the leader of a multidiscipline professional services firm, worked on civil engineering projects, and oversaw the company's efforts.

One of the firm's best clients is a department of transportation. When they conclude that our firm has the best skills, talent, and experience to perform the work they need, they ask us to prepare a work program of how many hours will be required by our talented team. Then they want to know how much we pay everyone on the team, and after that they audit our firm's financial information and tell us how much they will pay us, including the profit percentage they think is appropriate: hours times one, plus our audited overhead rate, plus their profit percent determines our fee.

How did we get this screwed up? We are selling *time*. Sure, it's the time of our team members and their rates of pay reflect their value in our firm and in the industry, but it still is selling *time*.

When we buy time from our accountant and attorney, they decide the rate. We may be buying time, but they decide the value of their time. Somehow that seems right. We engineers should also be determining the value of our services. Those services provide society with the infrastructure that it needs to operate, and our work lasts and contributes for decades—or in some cases hundreds of years.

So why in the world should we reduce the fee for our services if we could complete our tasks in less time? Are the results of our work not worth much more? Did we get ourselves into this dilemma because we are part of a process instead of the whole program? An attorney's advice ends upon receiving it; an accountant's too. But our efforts generally lead to a construction of facilities or a complex process that will be continued. Our efforts generally represent a small percentage of the cost of the final product, but that final product is essential for society to function. Therefore, our value should be significantly greater.

It is time we recognize that we design the systems that the entire planet runs on, walks on, drinks from, and—well, you get the point. We in the infrastructure design industry add value to the lives of people everywhere. Let's start charging for our value, not how efficient we are with our minutes. We got ourselves into this; we can get ourselves out. But it won't be easy, and it will take a significant amount of coordinated effort and some time. But hey, time is what we know about.

We do try hard to learn from others, both inside our industry and outside it. I think Jim Collins's *Good to Great* and *Built to Last* have had the biggest impact on our company so far. But many copies of David Maister's *True Professionalism* and *Practice What You Preach* are also purchased for some senior people. I am a big fan of Jerry Harvey, and if you haven't read *The Abilene Paradox* and my favorite, *How Come Every Time I Get Stabbed in the Back, My Fingerprints Are on the Knife?* you should read these just for the fun of them.

Jerry Harvey is the genius that proposes that at least 5 percent of your staff needs to be incompetent. Now, I know that might already be the case at most firms, but Jerry says that it is a requirement for high-performing firms. If 95 percent of your performers recognize that 5 percent of the staff is less-than-competent, the high performers will have no fear of layoffs. Plus, they recognize that this is a caring firm.

Taking the opposite position to General Electric CEO Jack Welch's policy of pruning the bottom 10 percent every year, the Jerry Harvey theory lets the production workers succeed without fear and forces up performance, driving up results. Jerry further proposes that he might open an employment agency to provide incompetent people to boost performance—if your firm should need some. You probably need to read the book.

I'm not saying DEA has 5 percent of its staff that are recognized as being incompetent, but by putting a lot of emphasis on helping all succeed and by demonstrating caring, we maximize employee loyalty and minimize turnover.

Chapter 12
People, People, People

"Don't imagine you're too familiar
and I don't see you anymore."
– Billy Joel

There are always a few things that make a firm unique. While some of our ideology is uniquely ours, there are also some unexpected events that have effects far beyond the expected.

Late in 1999, one of our secretarial staff found a kitten in our parking lot. It had no collar or ID indicating an owner. It was cold, hungry, and *cute*. "Can we keep it?"

While I am mildly allergic to cats, I have always loved them. Cats were banned from our home when our youngest son was determined to be highly allergic to them. The cat was "owned" by his older sister, who would have banned her brother instead of the cat, but cooler heads prevailed, and no cat has had a home with us since.

So when the kitten appeared on the office scene, a major management decision was called for from me. I said if the staff would care for the cat, be in on the weekends to check on it, feed it, handle the litter box, and so on, the cat could stay. My office, and the cat's home turf, was in the smallest of the four buildings the firm occupied at that time. So that office, an entire ten-thousand-square-foot building, became the domain of Millennium—the kitten's designation, as the turn of the millennium and the nonexistent "Y2K crisis" was imminent. Millennium became Millie, and she roamed at will within the office. This office building was old. When we purchased it, it had been an antique warehouse:

one big, open, sixteen-foot-high-ceilinged empty space.

As we made it into workspaces, no real offices were created; just a few walls went up here and there, but we added no doors. Of course, my space also had no door, and Millie soon decided that even though she had her own small cubicle nearby, my space was her favorite. And not just my space, but also my chair, my desk, and even my inbox were locations she could be found napping.  This story is not intended to establish how casual our office was, but having a wandering animal did set a tone.

The truly unexpected result of having Millie in my inbox occurred when we were trying to recruit a senior transportation-transit engineer from a large agency, where she had been the general manager of facilities. The process was not me interviewing her, but vice versa. Many very large firms were courting her, and she had visited several of them. David Evans and Associates was by far the smallest firm she was visiting.

When she came to our building, she was shown directly into my space. It seems that at other large firms, she was screened through several secretarial levels before entering the cloistered domain of the deciding leadership, so coming directly into my non-enclosed space was a different experience.

Far more different was that while I was trying to explain the wonders of our firm, Millie wandered across my desk and sat down for a nap in my inbox. I was accustomed to Millie, so I didn't give it much thought. Millie was quiet and not a distraction, I thought. After our conversation and my sharing of ideas on what this engineer might expect as a senior leader of the firm, she thanked me and left.

When she later called and said she would like to join the firm, I was extremely pleased and welcomed her to our team. On her arrival at our office, it became clear that she was a "cat person." Not only that, she had determined that any office that had a CEO with a cat in his inbox couldn't be too bad a place to work.

She (not Millie) has now led the firm on some of its largest-ever assignments. I guess I should award Millie at least a zero.

Human Resources or Herding Cats?

As I write about people, who are a consulting firm's only resource, I also gave this speech at a professional organization meeting:

"Managing in the New Millennium"

Human Resources from the CEO's Perspective

David Evans, CEO, David Evans and Associates, Inc.

It's our people!

When I wrote my outline, I started with "It's our people, stupid!" but the Human Resources (HR) department at DEA includes an editor, so you can see one of the duties of HR is to keep the CEO from writing things I probably shouldn't.

When I talk about DEA, it's David Evans and Associates, Inc., and not the other, more familiar federal agency. DEA is about 650 people located in seven states and a multidiscipline, twenty-office firm.

So HR, from my perspective, is mostly about supporting our widespread locations and promoting teamwork. Teamwork and team building always reminds me of my favorite managers and their great quotes. Yogi Berra, who said, "Nobody goes there anymore—it's too crowded" and K. C. Stengel, who described the secret to management as "keeping the people who hate you away from the people who are undecided." From my perspective, HR in the next millennium will be different from the past or today.

In 1961, when I was right out of college, jobs were scarce. When the dean said he had a job opening he knew about, I went to the company and prayed they would accept me. They looked at my transcript, looked at me, and said they had an entry-level position and when could I start? I said, "Now!" Since this was Thursday afternoon at about 4:00 p.m., they said Monday at 8:00 a.m. was fine and I was now employed! I had not asked what my salary would be, what benefits I would receive, or even what I might be doing.

Today, to capture a recent graduate, we must describe all of our compensation, benefits, vacation and sick leave policies and programs, plus describe not only the job, but the first project, before a person will judge us worthy of having them joining our team.

In the next millennium, the requirement to capture outstanding graduates will surely include all of the above, plus defining our commitment to technology, how virtual our offices really are—do they ever need to come to the office at all—and what kind of wrist-TV-Pentium-2000-GPS-locator-satellite-data-distribution system do we provide, and are we using AutoCAD 99 and Windows 2020. And, of course, how big is our signing bonus!

I think the major portion of the changes in our industry—both past and future—are from technology. Twenty years ago, when I started DEA with myself and one designer draftsman, HP had just created the calculator—the HP-35, and that meant we no longer needed trigonometry tables to calculate surveys and subdivisions. This was an incredible leap forward, which rendered Friden Calculators—the standard for the previous twenty-five years—obsolete and useless. As a CEO for a two-person firm, HR was making sure we had batteries for the HP-35, then 45, then 91, and coffee to keep both of us awake!

Years ago, when DEA had one hundred people, we had desktop HP 80 series computers with cassette tape programs as our standard, and we had an IBM Series I—with no software, but it had lots of blinking lights, which fascinated me. HR was now a payroll clerk, bookkeeper, and personnel records person—all in one. We had benefits: health insurance and vacation and sick leave, plus a retirement plan and all were handled by our comptroller.

So, as CEO, I focused on projects and business development and HR was a minor administrative effort with little attention from me.

Today—at 650 people—we have a four-person HR department that reports directly to the chief administrative officer (CAO).

In the next millennium, our strategic plan says we'll be well over one thousand people, and our HR department will include a technology officer—maybe a holographic one—to keep track of where the telecommuting—flextime—diverse and dispersed people are at any moment in time. And technology will continue to be the driving force as we become value-priced, information-based professionals instead of time-based service providers.

This is a little off the subject, but unless we get away from selling hours, technology will put us out of business. As we do more and more work with more and more technology in less and less time, we will be able to do our job in no time with very expensive equipment and give the work away. And we won't need HR people at all, as we will be out of business.

That brings me to the "where is the business/industry going"? part of my outline.

Many of you have heard of Dames and Moore and have heard of their acquisitions. In 1995 and 1996, Dames and Moore had acquired four firms and 1,500 new people and still planned to acquire businesses totaling over 200 million in revenue before the year 2000. It was later acquired by URS, now a firm of over 46,000. CH2M Hill has grown from 5,400 people in 1991, to 7,400 people in 1995 to over 30,000 employees in 2012. The big firms will get much bigger, and they will follow the global construction efforts. The infrastructure needs of the developing countries like China, which leads the world in needs for its booming, but fragile economy, is and will be inviting. But big firms—

the 10,000-plus employee firms—must work everywhere. They need fifteen to twenty million dollars of new projects every working day.

The other end of the spectrum is the specialty firm: a firm that does one thing exceptionally well and tries not to do anything else. These two types of firms are the ones most often mentioned as the survivors in the next millennium. HR work in them will be dramatically different. Big firms with incredible HR problems from multicountry operations will have 5 percent of its staff in HR-related efforts. And the small specialty firms will outsource HR to ... specialty firms in HR! I guess big firms may outsource HR to big HR firms. Maybe this is an opportunity!?

I am not a believer that the midsize firm won't survive. The key to survival for all firms—I believe—is client service and client relations. If there are enough clients for big firms, then there certainly are enough clients for the rest of us. If we can convince clients of the value we provide and create the trust that the client recognizes, then the size of the firm is not an issue.

It will require a large firm to provide all the needs of a large project or program. One way smaller to midsize firms can meet the needs of a major or nationwide client is by creating relationships with the megafirms, or relationships with other midsize firms, and by being leaders in technology.

At DEA, the HR manager's primary role is keeping us out of trouble and not only the trouble that our HR manager calls "alphabet soup"—ADA, OSHA, MWA, FLSA, FMLA, and NLRB—but unemployment departments, workers' compensation issues, and all new legal developments in the six states and one foreign country in which we operate.

Our HR manager has a law degree as well as a master's in HR studies. She views her team's goals as

- representing the corporate culture through corporate policies and programs,

- managing our record maintenance needs and ensuring documentation,
- managing our benefits program,
- participating in all our training programs—especially the ones relating to gender issues (we are still a male-dominated industry today in engineering and surveying), and
- ensure that we are a good place to work so we can attract and retain people.

When Anita Roddick, who founded The Body Shop chain, heard a CEO refer to people as his company's greatest asset, her comment was, "That's the most asinine comment I've ever heard! People aren't assets—they are the company!"

So keeping our human resources means keeping our company together—or "It's our people." Here is an example of being people focused: we hired a young graduate engineer, BSCE, who is deaf and cannot speak. We have provided the support and team members to work with him, and now, years later, he is a registered engineer on significant projects. His success is one of which we are very proud.

Well, I've reached the last items on my outline. Our industry and our company need more leaders: leaders to replace our retiring senior people and meet the needs of our clients, leaders in technology and business. To get them, we must make them: through training in our HR programs, project management training, writing training, listening training, sensitivity training, and leadership training. Those who are born leaders become leaders through experience and training.

Our expectation of our leaders is that they will recognize that people are our company and that human resources staff members are the primary caretakers of our people. If we train them to be sensitive, considerate, friendly, and patient, they can be leaders in this people business.

Chapter 13
Enjoyment Supports Creativity

"All I wanna do is have some fun. I got
a feeling I'm not the only one."
– Sheryl Crow

The roles that consulting engineers fulfill requires that significant creativity is being introduced, utilized, and interwoven into every design project. In DEA's effort to demonstrate we were a creative bunch in our earliest days, we began designing and printing a large (twenty-four by thirty-six inches) calendar to be given at Christmas/New Year to clients, friends, public agencies, and all the people we could think of. For it to be reasonably well received, it needed to be unique, playful, tasteful, and clever.

The first DEA calendar was for 1977, the first full year to come for us after our April 1, 1976, beginning. As the development boom was in full swing and every development project began with a mapping effort called the preliminary plat, which defined the location, size, and shape of each building site (lot), that was chosen as the theme of our first calendar. Thus, the 1977 calendar was a preliminary plat with a lot for each day and a block for each month.

Of course a preliminary plat includes cul-de-sacs, streets, lanes, boulevards, and open spaces, with each requiring a name. So the calendar had Tax Court in April; First Street in January; lanes named Lois, Frankie, and Jackla; a Mid-Year Boulevard; and a park space between October and November named Halloween Park.

Creating it was fun. Seeing our calendar on the walls of clients' offices—even public agency offices—was even more fun. Our effort was so well received that a new calendar had to be created every year. If an office had placed our two-foot-by-three-foot creation on a wall, taking it down meant an empty space needing a new calendar. The demand, which began with a printing of one hundred or so, grew by the end of 2005 into a print run of eight to ten thousand calendars. The time and effort to design a new creative masterpiece plus the printing—now in full color—made the expense grow into ten to twenty thousand dollars.

Achieving Zero

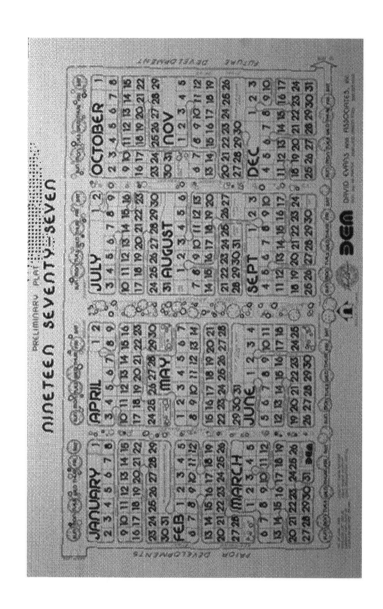

In September 2005, the time a calendar effort needed to be started for the 2006 edition, the disaster from Hurricane Katrina that had just occurred in August was being pictured every day in TV and newspapers. So, in addition to the 32,000 dollars donated by employees to disaster relief and a company match of 40,000, I announced the end of the calendar series: "We have donated the funds normally devoted to our annual calendar…to the relief effort.…I will miss the tradition of calendars, but as many have said, it is the right thing to do and doing what is right is a much better tradition."

The pictures below are from the calendars that grace the walls of our headquarters in Portland. I hope you can see some of the fun and creativity in these smaller representations.

Achieving Zero

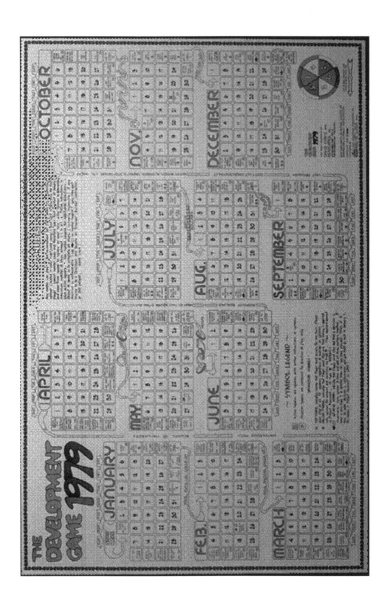

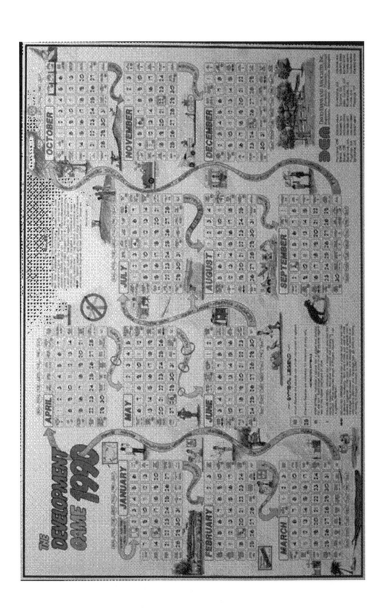

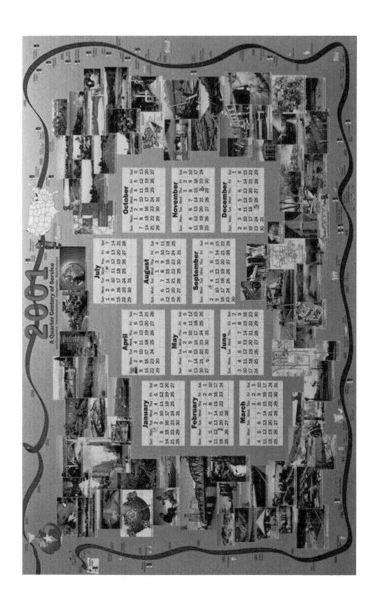

Chapter 14
Transitions

"How terribly strange to be seventy."
– Simon and Garfunkel

In my effort to describe the highs and lows, the wins and losses, the challenges and wonders of the consulting engineering practice and profession, I need to address the process of transitioning from founder, leader, CEO, and chairman to emeritus.

Having reached the role of chairman and installing a CEO who had been with the firm for twenty years, it was clear that the two of us, chairman and CEO, were both seeking to leave our responsibilities in about the same number of years.

The number we selected was seven. In seven more years, we would each leave the positions of authority in the hands of a younger generation.

While a seven-year plan may seem protracted, it was crafted to ensure that all the learning and wisdom we had acquired in our leadership roles was shared and passed on to the new team. Choosing the team to replace our roles as the "dynamic duo" became a topic of discussion and suggestions among the entire management group. The suggestions included the following:

Make a list of the top ten to twelve candidates, interview each, and then create a ranking system. The interviewers would be a committee from within the firm and maybe a member or two from the board of directors.

Look outside the firm for dynamic leaders from competing firms to enhance our position in the industry.

Name a search committee as the first step and have that committee determine the process.

And there were other suggestions as well.

After reviewing the suggestions, the CEO and I chose the two we wanted. No committees, no interviews, no discussions—we just decided.

That first decision was the easy part. We believed that creating a drawn-out selection process including many candidates would leave some unhappy, and those key persons who were ultimately not selected might feel dissatisfied and leave the firm. We also believed we already knew who would be the two candidates left standing, so why create a potentially unhappy situation?

Having made our selection and having announced the names of the two new leaders-to-be, we began a series of meetings as the Group of Four (chairman, CEO, two future leaders) to plan for the challenges ahead. The plan included conducting personality analyses, retaining a business planning consultant, setting mileposts for completing training, and setting goals for each of us in the program.

Our endeavor to pass on the learning we had absorbed in a lifetime of engineering and as the firm's chairman and CEO required that each of us take our new leadership duo to meet all the business, professional, and client contacts that we had established for ourselves. We then needed to determine at what point we would no longer attend the many industry and civic events where firm leaders make and maintain their contacts and when the new duo would step into those shoes as the firm's representatives. In many of these organizations and entities, we had significant friendships that would be hard to leave behind or hand off to our successors. But transitions mean change for all involved. The CEO and I were committed to seeing that the firm would not miss a beat in its business and professional life when we left. Our plan included an emergency process should a sudden death or disability occur during the transition.

As our seven-year plan proceeded, it did not exactly follow our designated course. Stuff happens.

One of the selected duo left the firm and began a firm of his own. This was a setback, but it made it even clearer to us that a transition program should not be hurried. A leadership transition program must be flexible enough to regroup along the way and continue. The learning, training, coaching, and introductions all needed to be completed with a new second person, but having been through it already, we now knew how to accomplish it.

In 2010, the next step for me was to move out of my office and have the new CEO take my former space. Moving out was physically simple. Stepping out from leadership roles after being the firm's founder was emotionally very difficult. The seven-year process eased the pain of leaving my creation (with lots of help from others) in new leadership hands. But no book can ever properly tell you how to let go of your children, your office, your firm, your creation. The last letter by me published in our in-house newsletter portrays how well our transition plan has worked. It surely rates a zero.

Chapter 15
MY Last Letter to the Firm

"You can check out anytime you want, but you can never leave."
– "Hotel California," The Eagles

Well, I have checked out of the many roles—founder, beginner, builder, sustainer, president, CEO, chairman—to now become a half-time chairman emeritus, a thirty-five year journey on which I am still traveling.

As I write this in my Santa Barbara office space, a new fiscal year for the firm is underway with a new CEO, with new leaders, 650-plus professionals in seventeen offices, all tackling the same issues that faced me thirty-five years ago. How do you find enough work, the right team, the honest clients, the best solutions to the hardest problems and then bill for your work, collect your bills, and make enough profit to do it all again and again and again? If you can do these things, you can stay in business for many, many years.

The business climate in 2009 and 2010 made the total of these challenges unattainable. We stayed in business, but we could not sustain a profit. Something more was necessary. We needed to learn from the hard years, change our systems, sharpen our focus, regain our sense of urgency, and endure the tough decisions that were necessary. If we could do all these things, we should be able to make 2011 a profitable fiscal year and ensure that DEA will be viable in 2012.

And we have.

Fiscal year 2011 ended with a profit, our accounts receivable are in the best condition in at least a decade, our backlog of work is significant, and our business units are poised for another improving year. It is not that the business climate has improved; it is that DEA has improved. We now know how to succeed in a hostile business environment. Our team is stronger and it has new, outstanding professionals doing what they do best.

All these things occurred by having the leadership group accept the challenge and diligently pursue the desired outcome.

Emeritus means retired but retaining a title. So I am semi-emeritus. As such, I claim no share of the successes achieved. I have offered my thoughts and opinions, but I know it takes a whole team to make a success, and the whole firm needed to work as one unit on the challenges to achieve the success. Without the diligent, caring, focused leadership DEA has in place today, we would not be ending 2011 with a significant profit and the feeling that 2012 will be even better.

What more could a chairman emeritus wish for? This founder, builder, etc., is very pleased and knows DEA is clearly on a path of continuing to improve the quality of life for our planet, our employees, and even for me!

Dave

Chapter 16
My Advice to Prospective Consulting Engineers

"What are you doing the rest of your life?
North and south and east and west of your life?"
– Alan and Marilyn Bergman

If this treatise has piqued your interest in the business of consulting engineering, and you would like to ask me a question or two, I'm available via e-mail at daveevansbook@gmail.com.

I have described a career that requires hard work and significant commitment, and is friendly and rewarding, not only in remuneration, but also significantly in pride of accomplishment.

So what advice can I give you if you embark on a life as a consulting engineer?

First, your college education is only the beginning of your education. Every project, task, study, and so on is an opportunity to learn. Keeping open to the wisdom of others and being discerning about truth are essential in your endeavors.

Second, value your close relationships in the industry and your business advisers. A mentor, teacher, colleague, or client can help you through many difficult days and tough choices. Always be open to advice and accept criticism.

Third, believe in yourself. When you believe you are right, be rigorous in your presentation. You may need to stop advocating, listen, and be appreciative of the other's positions. Once a conclusion is reached, whether you agree with it or not, accept the result and move on. Second-guessing, holding a grudge, and resenting decisions are never useful and will drain your energy and blur your focus.

Fourth, be a willing worker. An employee at all levels can and will be much more revered and appreciated if he or she takes on any task requested with a willing spirit and cheerful approach. And don't wait to be asked to do things you see should be done. If the garbage needs to go out, just do it!

Fifth, don't take yourself too seriously. Fun, enjoyment, and humor make all challenges less difficult and all problems more rewarding. The friendly business of consulting engineering will provide numerous opportunities for humor, enjoyment, and collaboration. Grasp as many as you can.

Sixth, keep your perspective on life. Remember why you are in the business and that it is only a way to provide you with the livelihood you seek. It is *not* your life; it is the way to achieve the life you want.

Seventh, be fearless in your entrepreneurial efforts. Wayne Gretzky says, "You miss 100 percent of the shots you don't take." And Yoda says, "Do! Or do not. There is no try." Not all of your great ideas will bring the results or rewards you hope for or expect, but the ideas you are tempted to disregard might turn out to be the best ideas you will have.

   Eighth, to be successful in the crazy business of consulting engineering, remember it is a business. While I hope I have reiterated some of the highs and lows, a brief summary follows:

- Know who you are and what values are important to you, and be sure all your colleagues and clients know as well.
- Know there are always wiser, more skilled, more intelligent, and more articulate individuals out there. Do your best to attract, hire, and retain all of them you can.
- Know that all your successes are the product of the collective efforts of you, your team, and your client, who brought you this problem to solve. Successes must be shared and enjoyed by all.
- Know that mistakes and errors are part of every life and all businesses. To err may be

human, but it is not fun for consulting engineers. When an error happens, seek only to fix or resolve the issue. Blame has no value and can cause lingering injuries. Learning from mistakes is the important result.
- Know that to be a leader you must have followers. They must know you and want to be with you. It is your tasks to make all followers achieve their own success, receive the recognition, and have each know you appreciate their support…and that the leader supports them.

Last, understand the reward for a completed task, a functioning project, a winning proposal, a miracle accomplished, and a finished assignment is that you did it. You said you could, you worked to the best of your talents, and the results are as you predicted.

Congratulations! You have achieved a zero. GO FOR IT!

Further Reading

Jim Collins. *Good to Great: Why Some Companies Make the Leap... and Others Don't.* New York: HarperCollins, 2001.

Jim Collins and Jerry Porras. *Built to Last: Successful Habits of Visionary Companies.* Harper Paperbacks, 2003.

Roger Connors, Tom Smith, and Craig Hickman. *The Oz Principle: Getting Results through Individual and Organizational Accountability.* Portfolio Hardcover; 2004.

Jerry B. Harvey. *The Abilene Paradox and Other Meditations on Management.* Jossey-Bass, 1988.

Jerry B. Harvey. *How Come Every Time I Get Stabbed in the Back My Fingerprints Are on the Knife? And Other Meditations on Management.* Jossey-Bass, 1999.

Song Lyrics and Excerpts

The Chapter	The Quote	The Source
Chapter One	"*The game commences, for the usual fee, plus expenses.*"	Song: "Private Investigations" Album: *Love Over Gold* Band: Dire Straits Year: 1982
Chapter Two	"*My, but we learn so slow, and heroes they come and they go.*"	Song: "Pretty Maids All In a Row" Album: *Hotel California* Band: The Eagles Year: 1976
Chapter Three	"*While the sorcerer slept the apprentice decided to play.*"	Song: "May Be a Price to Pay" Album: *The Turn of a Friendly Card* Band: Alan Parsons Project Year: 1980

Achieving Zero

The Chapter	The Quote	The Source
Chapter Four	"*Let me tell you a story. The Devil he has a plan.*"	Song: "Swamp" Album: *Speaking in Tongues* Band: Talking Heads Year: 1983
Chapter Five	"*If you could read my mind, love, what a tale my thoughts would tell.*"	Song: "If You Could Read My Mind" Album: *Sit Down, Young Stranger* Band: Gordon Lightfoot Year: 1970
Chapter Six	"*If you build it, they will come.*"	Movie: *Field of Dreams* Director: Phil Alden Robinson Character: The Voice Year: 1989 *This line is almost always misquoted in this way. The actual line is "If you build it, he will come," in reference to the spirit of baseball great, Shoeless Joe Jackson. But the misquote works better for this chapter.*

The Chapter	The Quote	The Source
Chapter Seven	"I've done some bad, I've done some good, I've done a whole lot better than they thought I would, so come on and treat me as you should."	Song: "Everybody Loves Me, Baby" Album: *American Pie* Band: Don McLean Year: 1971
Chapter Eight	"Brave Helios wake up your steeds, bring the warmth the countryside needs."	Song: "The Day Begins" Album: *Days of Future Passed* Band: The Moody Blues Year: 1967
Chapter Nine	"A man's got to know his limitations."	Movie: *Magnum Force* Director: Ted Post Character: Dirty Harry Callahan, played by Clint Eastwood Year: 1973
Chapter Ten	"Then they showed me a world where I could be so dependable, clinical, intellectual, cynical."	Song: "The Logical Song" Album: *Breakfast in America* Band: Supertramp Year: 1979

Achieving Zero

The Chapter	The Quote	The Source
Chapter Eleven	"The pattern of my life and the puzzle that is me."	Song: "Patterns" Album: *Parsley, Sage, Rosemary and Thyme* Band: Simon and Garfunkel Year: 1966
Chapter Twelve	"Don't imagine you're too familiar and I don't see you anymore."	Song: "Just The Way You Are" Album: *The Stranger* Band: Billy Joel Year: 1977
Chapter Thirteen	"All I wanna do is have some fun. I got a feeling I'm not the only one."	Song: "All I Wanna Do" Album: *Tuesday Night Music* Band: Sheryl Crow Year: 1993
Chapter Fourteen	"How terribly strange to be seventy."	Song: "Old Friends" Album: *Bookends* Band: Simon and Garfunkel Year: 1968
Chapter Fifteen	"You can check out anytime you want, but you can never leave."	Song: "Hotel California" Album: *Hotel California* Band: The Eagles Year: 1976

The Chapter	The Quote	The Source
Chapter Sixteen	*"What are you doing the rest of your life? North and south and east and west of your life?"*	Song: "What Are You Doing the Rest of Your Life?" Songwriters: Alan and Marilyn Bergman Year: 1969 *The Bergmans wrote this song for the 1969 motion picture* The Happy Ending. *It was nominated for an Academy Award for Best Original Song. Sarah Vaughn was awarded a Grammy for her 1973 recording, and thirty years later a second Grammy was awarded to Sting and Chris Botti for their version.*

About The Author

David F. Evans is the founder, chairman emeritus, and a member of the board of David Evans Enterprises, Inc., the holding company for David Evans and Associates (DEA), a multidisciplinary professional services firm headquartered in Portland, Oregon, that Dave started. Since its founding as a two-person land development design firm on April 1, 1976, DEA has become a recognized leader in the design and management of complex transportation, land development, water resources and energy projects nationwide. The February 4, 2008, issue of *Fortune* named the firm number seventy-three in the list of "100 Best US Companies to Work For" and DEA received one of the "2007 Oregon Ethics in Business" awards. The firm is ranked number eighty-seven in "Engineering News Record's April 2011 Top 500 Design" firms in the United States and number fifty-seven on the "Top 100 Pure Design Firms."

David Evans and Associates supports the efforts of seven hundred employee owners and maintains twenty-five offices throughout the United States. Current major projects include leading the design team for the four-billion-dollar new Columbia River I-5 Bridge and TriMet's 300-million-dollar Portland to Milwaukee LRT program. David Evans and Associates is also providing project management oversight for FTA on the Ferry Street and Battery Park Stations in New York City.

David F. Evans

A registered professional engineer and land surveyor by trade, Dave has professional licenses in Oregon, Washington, California, Hawaii, and New York and a bachelor's degree in civil engineering from Heald Engineering College. He also attended the Executive Program at Stanford Graduate School of Business. He is a fellow in the American Society of Civil Engineers (ASCE) and is a member of the American Council of Engineering Companies (ACEC) and ACEC of Oregon. Previously he was on the executive committee of the Design Professionals Coalition (DPC), and on the board of directors of the Construction Industry Round Table (CIRT). He is a current a board member of Mazzetti Nash Lipsey Burch (M+NLB) of San Francisco.

Dave, along with Ted Aadland, past president of AGC America, founded DEA's sister company Aadland Evans Constructors, Inc., a hard bid construction and CMGC company.

to do what they do best." DEA's growth has been based on that philosophy.

Dave led the leadership transition effort to prepare and install the next generation of DEA's leadership and management. The program was designed to be implemented over seven years and was complete in 2010 as Dave stepped down from the chairmanship.

Dave is a believer in ESOP plans and created DEA's plan in 1985 as the mechanism for ownership transition and an internal market for company stock. The ESOP now owns about 35 percent of the shares. All other outstanding shares are in the hands of employees as DEA's commitment to employee ownership continues.

Endnotes

1 Herbert Hoover, "Engineering as a Profession," *Engineer's Week*, 1954, http://www.hooverassociation.org/hoover/speeches/engineering_as_a_profession.php

2 Fraser S. Keith, originally published in *Engineering and Contracting*, May 22, 1918, p. 99.

Made in the USA
San Bernardino, CA
26 December 2013

Dave was honored to receive the SIR award from AGC of Oregon.

He is a past member of the board of the Japanese Garden Society of Oregon, the board of Oregon State Parks Trust, the Doernbecher Children's Hospital Foundation board, and the Oregon Public Broadcasting board.

Dave and his wife, Marcella, were married in 1959 and have three children and three grandchildren and now call Santa Barbara, California, home.

Dave's area of experience include the following:

Board of directors' governance as he was chairman for thirty-five years as the board grew from a three-person employee director's board to seven members with four being outside members to a nine-member board with five outside directors.

Dave guided the creation of DEA's values, vision, core purpose, and strategic direction, and he established the company philosophy: "We find outstanding professionals and give them freedom